AF501590

Series Editors

Advisory Board

Editors' Remarks

The idea of publishing high-level scientific manuscripts in a more informal way began in 1964 with **Lecture Notes in Mathematics**. In the meantime, more than 875 volumes of this series have appeared along with several other **Lecture Notes** series of different subjects, proving the outstanding acceptance of this type of publication by the scientific community. The term "Lecture Notes" can now be regarded as a Springer trademark.

The advantage of this informal, yet top-level publication is its short production time and comparatively low price. Authors are also free to publish their material in any other context after the appearance of a **Lecture Notes** volume. In addition, authors receive 75 complimentary copies of their contribution which they can disseminate to interested colleagues.

When Springer-Verlag approached us to discuss the need for a new **Lecture Notes** series in CONTROL and INFORMATION SCIENCES (LNCIS) it was rather easy for us to support this idea for two reasons. First, as already mentioned, the **Lecture Notes** series are regarded by leading scientists and engineers as a high-quality publication. Second, the fields of control and information science have become important areas of engineering and science in the last two decades. And nowadays these fields have a high degree of influence on our technological development.

Lecture Notes in Control and Information Sciences provide informal up to date reports in an area characterized by rapid theoretical development in conjunction with concrete practical applications. The topics covered include optimization, systems analysis, information and communication (including communication networks), and control systems.

Books in the series range from theoretical and/or practically oriented research or state of the art reports to conference proceedings.

The main goal of the publishers as well as the editors is a publication of superior stature which is in the interest of both authors and readers. General policy therefore requires that each volume be carefully considered before a decision about its acceptance is reached. Any questions and/or proposals sent to one of the series editors or to Springer-Verlag are always welcome. The success of a series like LECTURE NOTES IN CONTROL AND INFORMATION SCIENCES depends by and large on the interest and support of the scientific and engineering community. We hope that this new **Lecture Notes** series will interest the broad range of professionals concerned with control and/or information sciences.

A. V. Balakrishnan M. Thoma

Title Description

Volume 1

Distributed Parameter Systems: Modelling and Identification

Proceedings of the IFIP Working Conference, Rome, Italy, June 21–24, 1976

Editor: A. Ruberti
1978. 42 figures. V, 458 pages
DM 50,–; approx. US $ 23.30
ISBN 3-540-08405-3

Contents: Survey Papers: *A. V. Balakrishnan:* Identification of Distributed Parameter Systems: Non-Computational Aspects. – *J. L. Lions:* Some Aspects of Modelling Problems in Distributed Parameter Systems. – *J. H. Seinfeld, M. Koda:* Numerical Implementation of Distributed Parameter Filters with Application to Problems in Air Pollution. – **Contributed Papers:** *S. E. Aidarous:* On the Structure of the Control Subsystem for Stochastic Distributed Parameter Systems. – *S. E. Aidarous, M. R. Gevers, M. J. Installé:* On the Asymptotic Behavior of Sensors' Allocation Algorithm in Stochastic Distributed Systems. – *M. Amouroux, J. P. Babary, C. Malandrakis:* Optimal Location of Sensors for Linear Stochastic Distributed Parameter Systems. – *A. Bagchi:* Linear Smoothing in Hilbert Space. – *J. S. Baras:* Infinite Dimensional Filtering Problems in Optical Communication Systems. – *C. Bruni, G. Koch:* A Degenerate (Bounded Rate) Class of Distributed Parameter Systems. – *Z. Bubnicki, A. Kaczkowski, K. Nitka:* Mathematical Model and Identification of CO-Conversion Process. – *L. Carotenuto, G. di Pillo, G. Raiconi:* The Doubly Cubic Spline in the Identification of a Class of DPS, with Application to an Underground Aquifer. – *G. Chavent:* About the Identification and Modelling of Miscible or Immiscible Displacements in Porous Media. – *P. Colli Franzone, M. Stefanelli, C. Viganotti:* Identification of a Distributed Model for Ferrokinetics. – *R. F. Curtain, A. Ichikawa:* Optimal Location of Sensors for Filtering for Distributed Systems. – *M. C. Delfour, F. Trochu:* Discontinuous Finite Element Methods for the Approximation of Optimal Control Problems Governed by Hereditary Differential Systems. – *P. A. Fuhrmann:* On Spectral Minimality and Fine Structure of the Shift Realization. – *J. Henry:* Controllability of Some Nonlinear Parabolic Equations. – *R. Kluge, H. Langmach:* On Some Problem of Determination of Functional Parameter in Partial Differential Equations. – *M. Köhne:* Implementation of Distributed Parameter State Observers. – *K. Y. Lee:* Modeling and Estimation of Distributed Systems with Uncertain Parameters. – *N. Levan:* A State Space Realization of Linear Distributed Parameter System (DPS) Transfer Operators. – *A. J. Pritchard, E. P. Ryan:* Control and Identification of Distributed Parameter Systems. – *R. Triggiani:* On the Relationship Between First and Second Order Controllable Systems in Banach Spaces. – *S. G. Tzafestas:* Application of the Innovations Technique to Distributed-Parameter Detection and Estimation Problems. – *P. K. C. Wang:* Identification Problems in Plasma Physics. – *D. C. Washburn:* A Semigroup Theoretic Approach to Modeling of Boundary Input Problems.

Volume 2

New Trends in Systems Analysis

International Symposium, Versailles, France, December 13–17, 1976

Editors: A. Bensoussan, J. L. Lions
1977. 104 figures, 32 tables. VII, 759 pages (101 pages in French)
DM 68,–; approx. US $ 31.70
ISBN 3-540-08406-1

Contents: Table des matières: Control of Distributed Parameter Systems/Controle des Systèmes Distribués: *E. C. Hoppensteadt:* Optimal Exploitation of a Spatially Distributed Fishery. – *S. Tzafestas:* Distributed Parameter Nuclear Reactor Optimal Control. – *P. K. C. Wang:* On a Class of Optimization Problems Involving Domain Variations. – *B. v. d. Bosch:* Parameter Estimation in Distributed Chemical Systems. – *W. F. Ramirez, D. E. Clough:* On-Line Estimation and Identification of a Non-Linear, Distributed

Parameter Process: The Dehydrogenation of Ethylbenzene to Form Styrene in a Tubular, Fixed-Bed, Catalytic Reactor. – *Y. Kuroda, A. Makino:* Some Problems Arising in Distributed Parameter Reactor Systems. – *A. V. Balakrishnan:* Likelihood Ratios for Time-Continuous Data Models: The White Noise Approach. – *H. O. Fattorini:* Estimates for Sequences Biorthogonal to Certain Complex/Exponentials and Boundary Control of the Wave Equation. – *F. K. Greiss, W. H. Ray:* The Application of Distributed Parameter State Estimation Theory to a Metallurgical Casting Operation. – *M. Sorine:* Boucle ouverte et boucle fermée adaptée pour les systèmes distribués, un exemple d'application à la commande en temps réel d'un processus. – Industrial Robotics and Applications of Microprocessors/Robotique Industrielle et Applications des Microprocesseurs: *E. Freund, M. Syrbe:* Control of Industrial Robots by means of Microprocessors. – *J. L. Nevins, D. E. Whitney:* Categorization and Status of Assembly Research. – *C. J. Page, A. Pugh, W. B. Heginbotham:* Tactile Imaging for Component Recognition. – *L. J. Quagliata, Tze-Thong Chien, A. L. Hopkins Jr., J. Scott Rhodes:* Design and Analysis of Mass Production of Automotive Parts. – Systems Analysis in Problems of Energy/Application de L'Analyse de Systèmes aux Problèmes de L'Energie: *W. Häfele, R. Bürk, M. Breitenecker, C. Riedel:* Macro-Economic Models, Differential Topology and Energy Strategies. – *J. Weingart:* Systems Aspects of Large Scale Solar Energy Conversion. – *J. L. Abatut:* Analyse de systèmes et les problèmes posés par l'énergie solaire. – *H. R. Gruemm:* Resilience and its Application to Energy Systems. – *G. Kaufmann, E. Barouch:* Stochastic Modeling of Natural Resource Discovery – The Case of Oil and Gas. – *P. Courrège, J. M. Lasry:* Optimization du champ de miroirs d'une centrale solaire à concentation Ponctuelle. – Applications of Control Theory/Applications de la Théorie du Contrôle: *H. Kwakernaak:* Application of Control Theory to Population Policy. – *K. J. Aström:* Some Aspects on the Control of Large Tankers. – *J. Casti:* System Theory and Some of its Applications in Ecology, Water Resources and Energy. – *M. C. Delfour, A. Manitius:* Control Systems With Delays: Areas of Applications and Present Status of the Linear Theory. – *I. D. Landau, B. Courtiol, A. Fracon, L. Muller:* Applications de la Théorie du Contrôle dans les Aciéries. – *W. Findeisen:* Multilevel Structures for Control Systems. – *J. Karppinen, B. Blomsnes:* An Application of Optimization Methods to Spatial Control of Nuclear Reactor Cores. – *J. A. Bather:* Free Boundary Problems in Optimal Control. – *J. Zabczyk:* A Selection Problem Associated to a Renewal Process. – *E. Gelenbe:* Existence and Uniqueness of Stationary Distributions in a Model of Roll-Back-Recovery. – Control of Economic Systems/Contrôle des Systèmes Economiques: *M. Aoki:* Some Application of Control and System Theory in a Model of Dynamic Open Economy. – *N. Moisseev:* Systèmes cybernétiques et Problèmes de la gestion des processus économiques. – *M. Deleau, B. A. Oudet, P. Malgrange:* L'Application du contrôle aux modèles Macroéconomiques Francais: Expériences et perspectives d'avenir. – *J. H. Westcott:* An Experiment on Controlling a National Economy. – *K. D. Wall:* Time-Varying Models in Econometrics: Identifiability and Estimation. – *P. Nepomiastchy:* Méthodes d'optimization adaptées aux Modèles Macroéconomiques. – Environment and Pollution/Environnement et Pollution: *Y. Sawaragi, K. Inoue, H. Nakayama:* A Decision Making Model for Environmental Management Systems. – *A. Muramatsu, T. Watanabe, H. Akaike:* Integrated Model for Predicting the Regional Pollution for the Local Governments. – *J. Iisaka:* Environment Potential Survey by Remote Sensing. – *W. K. Foell, J. Buehring, W. Buehring, R. Dennis, K. Ito, R. Keeney, B. Lapillonne:* Long Term Policy Assessment of Energy/Environment Futures: A Systems Approach. – *G. Halbritter:* A Mathematical Model for Finding Compromises for Siting of Industrial Plants. – *D. M. Dubois:* On Temporal and Spatial Structure in Modes Systems and Application to Ecological Patchiness.

Volume 3

Differential Games and Applications

Proceedings of a Workshop,
Enschede, Netherlands,
March
16–25, 1977

Editors: P. Hagedorn,
H. W. Knobloch, G. J. Olsder
1977. 60 figures, 6 tables.
XII, 236 pages
DM 28,–; approx. US $ 13.10
ISBN 3-540-08407-X

Contents: *P. Bernhard:* Singular Surfaces in Differential Games: An Introduction. *A. Blaquiere:* Differential Games with Piece-Wise Continuous Trajectories. *J. V. Breakwell:* Zero-Sum Differential Games with Terminal Payoff. *J. Case:* Two Lectures on Cournot's Problem. *R. Elliott:* The Existence of Optimal Strategies and Saddle Points in Stochastic Differential Games. *R. Elliot:* Feedback Strategies in Deterministic Differential Games. *O. Hajek:* Toward a General Theory of Pursuit and Evasion. *G. Leitmann:* Many-Player Differential Games. *G. J. Olsder:* On Observation Costs and Information Structures in Stochastic Differential Games. *E. O. Roxin:* Differential Games with Partial Differential Equations. *T. L. Vincent:* Collision Avoidance at Sea. *T. L. Vincent:* Environmental Adaption by Annual Plants.

Volume 4

M. A. Crane, A. J. Lemoine

An Introduction to the Regenerative Method for Simulation Analysis

1977. 4 figures, 10 tables. VII, 111 pages

DM 18,50; approx. US $ 8.60

ISBN 3-540-08408-8

Contents: Basic Examples and Motivation. – The Regenerative Methods. – More Examples of Regenerative Processes. – The Regenerative Approach and Discrete – Event Simulations. – Approximation Techniques. – Alternative Ratio Estimators. – Some Other Results.

Many real-world problems involve stochastic systems. The most practical approach to their study is through simulation. Until recently, however, the statistical methodology available to simulators for analyzing the output of simulations of stochastic systems has been quite limited. A new technique of analysis called the regenerative method has recently been developed to deal with these problems. The method can produce valid and substantive statistical results for a very large and important set of simulations of systems with random elements. The goal of this book is to present the basic ideas and results of the regenerative method in a manner which may be easily understood by all potential users.

Volume 5

D. J. Clements, B. D. O. Anderson

Singular Optimal Control: The Linear-Quadratic Problem

1978. V, 93 pages

DM 18,50; approx. US $ 8.60

ISBN 3-540-08694-3

Contents: Singular Linear-Quadratic Optimal Control – A Broad Brush Perspective. – Robust Linear-Quadratic Minimization. – Linear-Quadratic Singular Control: Algorithms. – Discrete-Time Linear-Quadratic Singular Control and Constant Directions. – Open Questions.

This monograph is aimed at advanced graduate students, researchers and users of singular optimal control methods. It presumes prior exposure to the standard linear-quadratic regulator problem, and a general maturity in linear systems theory. A number of advances in singular, linear-quadratic control have taken place very recently. The book is intended to present an up-to-date account of many of these. At the same time, the book presents a unified view of various approaches to singular optimal control, many of which are apparently unrelated.

Volume 6

Optimization Techniques

Proceedings of the 8th IFIP Conference on Optimization Techniques, Würzburg, September 5–9, 1977

Part 1

Editor: J. Stoer

1978. 115 figures. XIII, 528 pages

DM 55,–; approx. US $ 25.60

ISBN 3-540-08707-9

Contents: Invited Speakers: *A. V. Balakrishnan:* Stochastic Optimization: Time-Continuous Data Models. *M. R. Hestenes:* Conjugate Direction Methods in Optimization. *J. L. Lions:*

Remarks on the Relationships between Free Surfaces and Optimal Control of Distributed Systems. *G. I. Marchuk:* Some Mathematical Models in Immunology. *H. J. Sussmann:* On Some Self-Immunization Mechanisms of Applied Mathematics: The Case of Catastrophe Theory. – **Round Table Discussion on World Models:** *D. C. J. de Jongh:* World Models. *H. D. Scolnik:* World Modeling. *R. Tomović:* Limitations of World Models. – **Computational Techniques in Optimal Control:** *M. C. Bartholomew-Biggs:* The Use of Nonlinear Programming in a Direct/Indirect Method for Optimal Control Problems. *F. Kappel:* Approximation of Functional-Differential Equations by Ordinary Differential Equations and Hereditary Control Problems. *R. Gonzalez, E. Rofman:* An Algorithm to Obtain the Maximum Solution of the Hamilton-Jacobi Equation. *M. P. J. Scott, A. A. Dickie:* Time Optimal Control of State Constrained Linear Discrete Systems. – **Stochastic Optimal Control:** *A. Benveniste, M. Goursat, G. Ruget:* A Robust Adaptive Procedure for Solving a Non Gaussian Identification Problem. *R. Drenick:* Optimization and Uncertainty. *A. Bagchi, H. Kwakernaak:* The Separation Principle for the Control of Linear Stochastic Systems with Arbitrary Information Structure. *S. Maurin:* A Decomposition Scheme for the Hamilton-Jacobi Equation. *D. J. Mellefont, R. W. H. Sargent:* Calculation of Optimal Measurement Policies for Feedback Control of Linear Stochastic Systems. *U. Pursiheimo:* On the Optimal Search for a Moving Target in Discrete Space. *M. Robin:* Optimal Maintenance and Inspection: An Impulsive Control Approach. *L. Socha, J. Skrzypek:* Application of Open Loop Control to the Determination of Optimal Temperature Profile in the Chemical Reactor. *Gy. Sonnevend:* Output Regulation in Partially Observable Linear Disturbed Systems. *A. Waberski:* Reaction of Continuous Dynamic Systems with Complex form under Time-Space Random Fields. – **Differential Games:** *J. L. Goffin, A. Haurie:* On the Characterization and the Computation of the Characteristic Function of a Game without Side Payments. *G. Leitmann, H. S. Liu:* Evasion in the Plane. *G. J. Olsder, J. L. Walter:* A Differential Game Approach to Collision Avoidance of Ships. *B. Tolwinski:* A Method for Computing Nash Equilibria for Non-Zero-Sum Differential Games. – **Optimal Control of Partial Differential Equations:** *G. Chavent, G. Cohen:* Numerical Approximation and Identification in a 1-D Parabolic Degenerated Non-Linear Diffusion and Transport Equation. *J. Eichler:* Optimization of the Design of an In-Flight Refueling System. *T. Futagami:* The FE and LP Method and the Related Methods for Optimization of Partial Differential Equation Systems. *R. Klötzler:* A Generalization of the Duality in Optimal Control and Some Numerical Conclusions. *W. Krabs:* On Optimal Damping of One-Dimensional Vibrating Systems. *N. Levan, L. Rigby:* Stability and Stabilizability of Linear Control Systems on HILBERT Space via Operator Dilation Theory. *I. Lasiecka, K. Malanowski:* On Discrete-Time Ritz-Galerkin Approximation of Control Constrained Optimal Control Problems for Parabolic Equations. *B. Rousselet:* Optimal Design and Eigenvalue Problems. *E. Sachs:* Optimal Control of Parabolic Boundary Value Problem. *C. Saguez:* A Variational Inequality Associated with a Stefan Problem Simulation and Control. *C. L. Simionescu:* Numerical Methods for a Generalized Optimal Control Problem. *R. Triggiani:* A Cosine Operator Approach to Modelling Boundary Input Hyperbolic Systems. *G. Volpi, P. Sguazzero:* The Linearization of the Quadratic Resistance Term in the Equations of Motion for a Pure Harmonic Tide in a Canal and the Identification of the CHEZY Parameter C. – **Immunology, Disease and Control Theory:** *D. Dubois, G. Montfort:* Stochastic Simulation of Space-Time Dependent Predator-Prey Models. *C. Bruni, A. Germani, G. Koch:* Optimal Derivation of Antibody Distribution in the Immune Response from Noisy Data. *R. R. Mohler, C. F. Barton:* Compartmental Control Model of the Immune Process. *A. S. Perelson:* The IgM-IgG Switch Looked at from a Control Theoretic Viewpoint. – **Environmental and Energy Systems:** *M. Bielli, G. Calicchio, M. Cini, F. Nicolo:* Two-Level Optimization Techniques in Electric Power Systems. *G. Halbritter:* Multiobjective Programming and Siting of Industrial Plants. *P. G. Harhammer:* Economic Operation of Electric Power System under Environmental Impacts. *I. Nakahori, I. Sakaguchi, J. Ozawa:* An Optimum Operation of Pump and Reservoir in Water Supply System. *K. Ogino:* Optimal Expansion of Generating Capacity in National Electric Power Energy System. *A. G. Ulusoy, D. M. Miller:* Pipeline Network Optimization – An Application to Slurry Pipelines. *L. F. Escudero, A. M. Vazquez-Muniz:* Linear Fitting of Non-Linear Functions in Optimization. A Case Study: Air Pollution Problems.

Volume 7

Optimization Techniques

Proceedings of the 8th IFIP Conference on Optimization Techniques, Würzburg, September 5–9, 1977

Part 2

Editor: J. Stoer
1978. 82 figures. XIII, 512 pages
DM 55,–; approx. US $ 25.60
ISBN 3-540-08708-7

Contents: Mathematical Programming, Theory: *R. P. Hettich, H. Th. Jongen:* Semi-Infinite Programming: Conditions of Optimality and Applications. *T. Zolezzi:* On Equiwellset Minimum Problems. *H. Maurer, J. Zowe:* Second-Order Necessary and Sufficient Optimality Conditions for Infinite-Dimensional Programming Problems. – **Nonlinear and Stochastic Programming:** *P. Deuflhard, V. Apostolescu:* An Underrelaxed Gauss-Newton Method for Equality Constrained Nonlinear Least Squares Problems. *J. Bräuninger:* A Modification of Robinson's Algorithm for General Nonlinear Programming Problems Requiring only Approximate Solutions of Subproblems with Linear Equality Constraints. *U. Eckhardt:* On a Minimization Problem in Structural Mechanics. *J. N. Holt:* Non-Linear Least Squares Inversion of an Integral Equation Using Free-Knot Cubic Splines. *S. Bali, S. E. Jacobsen:* On the Convergence of the Modified TUI Algorithm for Minimizing a Concave Function on a Bounded Convex Polyhedron. *B. D. Kiekebusch-Müller:* A Class of Algorithms for the Determination of a Solution of a System of Nonlinear Equations. *K. Marti:* Stochastic Linear Programs with Random Data Having Stable Distributions. *G. G. L. Meyer:* Methods of Feasible Directions with Increased Gradient Memory. *P. Mitter, C. W. Ueberhuber:* The Continuous Method of Steepest Descent and its Discretizations. *V. H. Nguyen, J. J. Strodiot:* Convergence Rate Results for a Penalty Function Method. *S. S. Oren:* A Combined Variable Metric – Conjugate Gradient Algorithm for a Class of Large Scale Unconstrained Minimization Problems. *V. E. Krivonozhko, A. I. Propoi:* Simplex Method for Dynamic Linear Program Solution. *K. Schittkowski:* An Adaptive Precision Method for the Numerical Solution of Constrained Optimization Problems Applied to a Time-Optimal Heating Process. *A. Friedlander, J. M. Martinez, H. D. Scolnik:* Generalized Inverses and a New Stable Secant Type Minimization Algorithm. *F. Sloboda:* A Conjugate Directions Method and its Application. *R. W. H. Sargent, G. R. Sullivan:* The Development of an Efficient Optimal Control Package. *J. Szymanowski, A. Ruszczynski:* An Accuracy Selection Algorithm for the Modified Gradient Projection Method in Minimax Problems. *A. Wierzbicki, A. Janiak, T. Kreglewski:* Single-Iterative Saddle-Point Algorithm for Solving Generally Constrained Optimization Problems Via Augmented LAGRANGEans. – **Integer Programming, Networks:** *G. d'Atri:* Improved Lower Bounds to 0/1 Problems via LAGRANGEan Relaxation. *G. Finke:* A Unified Approach to Reshipment, Overshipment and Post-Optimization Problems. *Hai Hoc Hoang:* Solving an Integer Programming Problem. *D. Hausmann, B. Korte:* Worst Case Analysis for a Class of Combinatorial Optimization Algorithms. *L. Mihalyffy:* An Improved Method of Successive Optima for the Assignment Problem. *M. Minoux:* Accelerated Greedy Algorithms for Maximizing Submodular Set Functions. *R. Petrovic:* Resource Allocation in a Set of Networks under Multiple Objectives. *Z. Pogany:* An Algorithm for Solving the Generalized Transportation Problem. *T. M. Simundich:* An Efficient Algorithm for Solving a Stochastic, Integer Programming Problem Arising in Radio Navigation. *S. Walukiewicz:* Using Pseudoboolean Programming in Decomposition Method. *R. Slowinski, J. Weglarz:* Solving the General Project Scheduling Problem with Multiple Constrained Resources by Mathematical Programming. *U. Zimmermann:* Threshold Methods for Boolean Optimization Problems with Separable Objectives. – **Urban Systems:** *A. Lukka:* Comparison of some Educational Planning Models. *R. Minciardi, P. P. Puliafito, R. Zoppoli:* Mathematical Programming in Health-Care Planning. *M. L. Costa Lobo, L. Valadares Tavares, R. Carvalho Oliveira:* A Model of Housing Developing Costs Related to Location. *A. Cumani, R. Del Bello, A. Villa:* An Optimum Surveillance and Control System for Synchronized Traffic Signals. *K. Yajima:* Regional Classification Problem and Weaver's Method. *P. Vicentini, B. Zanon:* A Mathematical Model for Decision Making in Public Service Planning. – **Economics:** *M. T. Hilhorst, G. J. Olsder, C. W. Strijbos:* Optimal Control of Regional Economic Growth. *H. Myoken, Y. Uchida:* System Modeling for Interconnected Dynamic Economy and the Decentralized Optimal Control. *M. Luptacik:* Economic Consequences of a Change in Demographic Patterns: A Linear Programming Model. – **Operations Research:**

F. Archetti, B. Betro: The Multiple Covering Problem and its Application to the Dimensioning of a Large Scale Seismic Network. *M. Cirina:* A Remark on Econometric Modelling, Optimization and Decision Making. *J. Grabowski:* Formulation and Solution of the Sequencing Problem with Parallel Machines. *A. Jakubowski:* Stochastic Model of Resource Allocation to R and D Activities under Cost Value Uncertainty. *C. Colding-Jørgensen, O. H. Jensen, P. Stig-Nielsen:* Scheduling of Trains – An Optimization Approach. *B. L. Miller:* Optimal Portfolios where Proceeds are a Function of the Current Asset Price. *L. Bianco, B. Nicoletti, S. Ricciardelli:* An Algorithm for Optimal Sequencing of Aircraft in the Near Terminal Area. – **Computer and Communication Networks, Software Problems:** *B. Camoin:* A Mathematical Model of Traffic in Communication Networks. *J. Kacprzyk, W. Stanczak:* Partitioning a Computer Network into Subnetworks and Allocation of Distributed Data Bases. *H. Kondo, I. Yoshida, T. Kato:* Effective File Allocation Method onto Disc Devices. *G. Mazzarol, E. Tomasin:* Optimal File Allocation Problem and Relational Distributed Data Bases. *D. F. Rufer:* General Purpose Nonlinear Programming Package.

Volume 8
R. F. Curtain, A. J. Pritchard

Infinite Dimensional Linear Systems Theory

1978. VII, 297 pages
DM 36,50; approx. US $ 17.00
ISBN 3-540-08961-6

Contents: Semigroup Theory. – Controllability, Observability, and Stability. – Quadratic Cost Control Problem. – Stochastic Processes and Stochastic Differential Equations. – The State Estimation Problem. – The Separation Principle for Stochastic Optimal Control. – Unbounded Control and Sensing in Distributed Systems. – Time Dependent Systems.

The philosophy underlying this book is to develop a mathematical framework which enables the generalization of the finite dimensional results to infinite dimensions and which includes both distributed parameter systems and differential delay systems as special cases. It describes the system dynamics in terms of a strongly continuous semigroup on an appropriate Banach space. Using this unifying mathematical approach it is possible to clarify the essential concepts of observability, controllability, the quadratic cost control problem, and the estimation and control problems for stochastic systems. The authors have further examined the implications of the abstract theory to specific examples of distributed and delay systems.

Volume 9
Y. M. EL-Fattah, C. Foulard

Learning Systems: Decision, Simulation, and Control

1978. 27 figures, 2 tables. VII, 119 pages.
DM 18,50; approx. US $ 8.60
ISBN 3-540-09003-7

Contents: Abstract. – Cybernetics of Learning. – Decision – Pattern Recognition. – Simulation – Models of Collective Behavior. – Control – Finite Markov Chains. – Epilogue.

This monograph studies the use of learning systems for decision, simulation, and control. Chapter I discusses what is meant by learning systems and comments on their cybernetic modeling. Chapter II, which concerns decision, is devoted to the problem of pattern recognition. Chapter III deals with simulation and focuses on a certain class of collective behavior problems. Chapter IV is on control and presents a simple model of finite Markov chains. For each of the last three chapters numerical examples are worked out using only computer simulations. Mathematicians and engineers in research institutes and at universities will appreciate this valuable survey.

Volume 10
J. M. Maciejowski

The Modelling of Systems with Small Observation Sets

1978. 9 figures, 6 tables. VI, 242 pages
DM 28,–; approx. US $ 13.10
ISBN 3-540-09004-5

Contents: Introduction. – Survey of Related Work. – A Characterisation of Modelling. – Incorporation of A Priori Knowledge. – Fragments of Programming Languages. - λ-Comparability. – Table Look-Up Codings. – Discussion and Conclusion. – Appendices: A: Formal Semantics of Programming Languages. B: Syntax of the Algol W-Support of the Gas-Furnace Models. C: Table Look-Ups for the Gas-Furnace Models. Diagrams.

The book introduces and discusses the problem of assessing and interpreting models of systems when only small sets of observations are available. As a general criterion of the quality of a model, the concept of Information Gain is proposed. It is shown that information gain is a suitable criterion for a wide class of models, including nonlinear dynamical stochastic models, and that its computation is straightforward. Its use for the assessment of rival models is demonstrated. Finally the effect system observations coding has on model assessment is examined.

Volume 11
Y. Sawaragi, T. Soeda,
S. Omatu

Modeling, Estimation, and Their Applications for Distributed Parameter Systems

1978. 4 figures. VI, 269 pages
DM 32,50; approx. US $ 15.20
ISBN 3-540-09142-4

Contents: Introduction. – Mathematical Preliminaries. – Optimal Estimation Problems for a Distributed Parameter System. – Existence Theorems for the Optimal Estimations. – Optimal Sensor Location Problems. – Stochastic Optimal Control Problems.

Two important problems in technology and applied science are estimation and optimal control of the behavior of physical processes subject to random disturbances and observation errors. This book deals with a class of problems within the framework of a general problem of estimation and control for a linear distributed parameter system. The estimation problems are those of obtaining an optimum approximation to the time history of a process'es behavior from noisy observation data. The control problem is that of determining inputs to a process to achieve desired goal such as maximum yield or minimum expenditure of fuel in spite of the random disturbances present.

Volume 12
I. Postlethwaite,
A. G. J. Macfarlane

A Complex Variable Approach to the Analysis of Linear Multivariable Feedback Systems

1979. 53 figures. IV, 177 pages.
DM 21,50; approx. US $ 10.00
ISBN 3-540-09340-0

Contents: Introduction. – Preliminaries. – Characteristic Gain Functions and Characteristic Frequency Functions. – A Generalized Nyquist Stability Criterion. – A Generalized Inverse Nyquist Stability Criterion. – Multivariable Root Loci. – On Parametric Stability and Future Research. – Appendices. – References. – Bibliography. – Index.

This text extends the concepts underlying the techniques of Nyquist, Bode and Evans to multivariable systems. It is shown how studying complex gain as a function of complex frequency, and vice versa, can be extended to the multivariable case by associating a pair of analytic functions with transfer function matrices: a characteristic gain and a characteristic frequency function. The generalized Nyquist stability criterion for multivariable feedback systems is presented in Chapter 4. In Chapter 5 a generalization of the inverse Nyquist stability criterion to the multivariable case is developed. The Evans' root locus approach is extended to multivariable systems in Chapter 6. The effect of parameter variations on a multivariable feedback system is considered in Chapter 7 by introducing the concepts of "parametric" root loci and "parametric" Nyquist loci.

Volume 13
E. D. Sontag

Polynomial Response Maps

1979. VIII, 168 pages.
DM 23,–; approx. US $ 10.70
ISBN 3-540-09393-1

Contents: Introduction. – Algebraic Preliminaries. – Realization Theory. – Finiteness Conditions. – State-Affine Systems. – Classes of Quasi-Reachable Realizations. – Other Topics. – References. – Glossary of Notations. – Index.

This work introduces a new class of nonlinear response maps, called polynomial maps. The natural internal (state-space) realizations of polynomial response maps are studied and a result is obtained on the existence and uniqueness of canonical realizations. Various external finiteness conditions are related to corresponding ones of internal realizations. An investigation is made, in particular, into the existence of input/output equations. Algebra and geometry are the main tools used. The subclass of bounded polynomial response maps is also introduced, enabling the derivation of many realization results from automata and language theory.

Volume 14

International Symposium on Systems Optimization and Analysis

Rocquencourt, December 11–13, 1976
IRIA LABORIA
Institut de Recherche d'Informatique et d'Automatique, Rocquencourt-France
Editors: A. Bensoussan, J. L. Lions
1979. 16 figures, 10 tables. VIII, 332 pages (30 pages in French).
DM 36,50; approx. US $ 17.00
ISBN 3-540-09447-4

Contents: Economical Models/Modèles Economiques: *M. A. Keyzer:* An International Agreement as a Complementarity Problem. *L. S. Lasdon, A. Meeraus:* Solving Nonlinear Economic Planning Models Using GRG Algorithms. *K. D. Wall:* Specification and Estimation of Econometric Models with Generalized Expectations. *A. Drud:* Implementation of the Model in Codes for Control of Large Econometric Models. *P. Nepomiastchy, B. Oudet, F. Rechenmann:* MODULECO, aide à la construction et à l'utilisation de modèles macroéconomiques. – **Identification, Estimation, Filtering/Identification, Estimation, Filtrage:** *E. Wong:* A Calculus of Multiparameter Martingales and Its Applications. *T. Kailath, A. Vieira, M. Morf:* Orthogonal Transformation (Square Root). Implementations of the Generalized Chandrasekhar and Generalized Levinson Algorithms. *J. Rissanen:* Shortest Data Description and Consistency of Order Estimates in Arma-Processes. *E. A. Jonckheere, L. M. Silverman:* Spectral Theory of Linear Control and Estimation Problems. *M. Clerget, F. Germain:* Un algorithme de lissage. *H. K. Khalil, B. F. Gardner Jr., J. B. Cruz Jr., P. V. Kokotovic:* Reduced Order Modeling of Closed-Loop Nash Games. *S. K. Mitter, S. K. Young:* Quantum Estimation Theory. – **Adaptive Control/Contrôle Adaptatif:** *K. J. Aström:* Piece-Wise Deterministic Signals. *V. Borkar, P. Varaiya:* Adaptive Control of Markov Chains. *Y. C. Ho, R. Suri:* Resource Management in an Automated Warehouse. *Y. Landau:* Dualité asymptotique entre les systèmes de commande adaptative avec modèle et les régulateurs à variance minimale auto-ajustables. – **Numerical Methods in Optimization/Méthodes Numériques en Optimization:** *B. T. Poljak:* On the Bertsekas' Method for Minimization of Composite Functions. *E. A. Nurminski:* On ε-Subgradient Methods of Non-Differentiable Optimization. *J. F. Shapiro:* Non-Differentiable Optimization and Large Scale Linear Programming. *D. P. Bertsekas:* Algorithms for Non-Linear Multicommodity Network Flow Problems. *J. Hald, K. Madsen:* A 2-Stage Algorithm for Minimax Optimization. – **Distributed Systems/Systèmes Distribués:** *A. G. Butkovskiy:* Certain Control Problems in Distributed Systems. *D. G. Lainiotis:* Partitioning: The Multi-Model Framework for Estimation and Control. *D. L. Russel, R. M. Reid:* Water Waves and Problems of Infinite Time Control. *R. Triggiani:* Boundary Stabilizability for Diffusion Processes. *H. T. Banks, J. A. Burns, E. M. Cliff:* Spline Based Approximation Methods for Control and Identification of Hereditary Systems. *J. Zabczyk:* Stabilization of Boundary Control Systems.

Volume 15

Semi-Infinite Programming

Proceedings of a Workshop, Bad Honnef, August 30 – September 1, 1978
Editor: R. Hettich
1979. 9 figures, 9 tables. X, 178 pages
DM 23,–; approx. US $ 10.70
ISBN 3-540-09479-2

Contents: Theory: *K. Glashoff:* Duality Theory of Semi-Infinite Programming. *A. Ben-Tal, M. Teboulle, J. Zowe:* Second Order Necessary Optimality Conditions for Semi-Infinite Programming Problems. *H.-J. Kornstaedt:* Necessary Conditions of Higher Order for Semi-Infinite Programming. – **Methods for Linear Problems:** *S.-A. Gustafson:* On Numerical Analysis in Semi-Infinite Programming. *P. R. Gribik:* A Central-Cutting-Plane Algorithm for Semi-Infinite Programming Problems. *K. Roleff:* A Stable Multiple Exchange Algorithm for Linear SIP. – **Methods for Nonlinear Problems:** *R. Hettich, W. van Honstede:* On Quadratically Convergent Methods for Semi-Infinite Programming. *R. Hettich:* A Comparison of some Numerical Methods for Semi-Infinite Programming. *W. van Honstede:* An Approximation Method for Semi-Infinite Problems. **Applications:** *S.-A. Gustafson:* On Semi-Infinite Programming in Numerical Analysis. *G. Dahlquist, G. H. Golub, St. G. Nash:* Bounds for the Error in Linear Systems. *W. Krabs:* One Sided L_1-Approximation as a Problem of Semi-Infinite Linear Programming.

Volume 16

Stochastic Control Theory and Stochastic Differential Systems

Proceedings of a Workshop of the "Sonderforschungsbereich 72 der Deutschen Forschungsgemeinschaft an der Universität Bonn" which took place in January 1979 at Bad Honnef
Editors: M. Kohlmann, W. Vogel
1979. 15 figures, 1 table. XII, 615 pages (23 pages in French).
DM 65,–; approx. US $ 30.30
ISBN 3-540-09480-6

Contents: Part I: SURVEY LECTURES. *A. V. Balakrishnan:* White Noise Models in Non-Linear Filtering and Control. *A. Bensoussan:* Optimal Impulsive Control Theory. *J. M. Bismut:* An Introduction to Duality in Random Mechanics. *R. F. Curtain:* Linear Stochastic Ito Equations in Hilbert Spaces. *M. H. A. Davis:* Martingale Methods in Stochastic Control. *T. E. Duncan:* A Geometric Approach to Linear Control and Estimation. *R. J. Elliott:* The Martingale Calculus and Applications. *A. Friedman:* Interaction Between Stochastic Differential Equations and Partial Differential Equations. *H. Kushner:* Approximation of Solutions to Differential Equations with Random Inputs by Diffusion Processes. *R. Rishel:* Optimal Conditions and Sufficient Statistics for Controlled Jump Processes. *J. van Schuppen:* Stochastic Filtering Theory: A Discussion of Concepts, Methods, and Results. *J. Zabczyk:* Introduction to the Theory of Optimal Stopping. **Part II:** RESEARCH REPORTS. *A. Al-Hussaini, R. J. Elliott:* Weak Martingales Associated with a TwoParameter Jump Process. *T. Basar:* Stochastic Stagewise Stackleberg Strategies for Linear Quadratic Systems. *E. O. Bertsch:* Some Remarks Concerning Attainable Sets of Stochastic Optimal Control. *J. M. Bismut:* Potential Theory in Optimal Stopping and Alternating Processes. *V. Borkar, P. Varaiya:* Adapted Control of Markov Chains. *N. Christopeit:* Solution of the Limited Risk Problem Without Rank Conditions. *M. Deistler:* The Parameterization of Rational Transferfunction Linear Systems. *G. de Mey:* A Stochastic Model for the Electrical Conduction in Non Homogenous Layers. *B. Doshi:* Policy Improvement Algorithm for Continuous Time Markov Decision Processes with Switching Costs. *T. E. Duncan:* An Algebro-Geometric Approach to Estimation and Stochastic Control for Linear Pure Delay Time Systems. *T. Eisele:* A Non-Linear Martingale Problem. *A. Ferroni, G. S. Goodman, A. Moro:* Pathwise Construction of Random Variables and Function Space Integrals. *T. Gasser, G. Dummermuth:* Non-Gaussianity and Non-Linearity in Electroencephalographic Time Series. *B. Grigelionis:* Canonical Form and Local Characteristics of Semimartingales. *M. Hazewinkel:* On Identification and the Geometry of the Space of Linear Systems. *K. Helmes:* A Numerical Comparison of Non Linear with Linear Prediction for the Transformed Ornstein Uhlenbeck Process. *U. Herkenrath, R. Theodurescu:* On the Bandit Problem. *J. Jacod:* Existence and Uniqueness for Stochastic Differential Equations. *A. Kistner:* On the Solution and the Moments of Linear Systems with Randomly Disturbed Parameters. *W. Kliemann:*

Some Exact Results on Stability and Growth of Linear Parameter Excited Stochastic Systems. *M. Kohlmann, R. Rishel:* A Variational Inequality for a Partially Observed Stopping Time Problem. *H. Korezlioglu, G. Mazziotto, J. Szpirglas:* Equations du Filtrage Non Linéaire pour des Processus a Deux Indices. *A. J. Krener:* Minimum Covariance, Minimax and Minimum Energy Linear Estimators. *H. Kunita:* Non Linear Filtering for the System with General Noise. *E. Pardoux:* Filtering of a Diffusion Process with Poisson Type Observation. *D. Plachky:* On Weak Closures of Convex and Solid Sets of Probability Measures. *M. M. Rao:* Non L^1-Bounded Martingales. *B. Rustem, K. Velupillai:* On the Definition and Detection of Structural Change. *G. Sawitzki:* Exact Filtering in Exponential Families: Discrete Time. *A. Segall:* Lower Estimation Error Bounds for Gauss-Poisson Processes. *R. Sentis:* Sur L'Approximation d'un Processus de Transport par une Diffusion. *S. E. Shreve:* Resolution of Measurability Problems in Discrete-Time Stochastic Control. *R. Tarres:* Optimal Non-Explosive Control of a Non Constrained Diffusion and Behaviour when the Discount Vanishes. *H. Walk:* Sequential Estimation of the Solution of an Integral Equation in Filtering Theory. *M. P. Yershov:* Causal and Non-Anticipating Solutions of Stochastic Equations.

Volume 17

O. I. Franksen, P. Falster, F. J. Evans

Qualitative Aspects of Large Scale Systems

Developing Design Rules Using APL

1979. XII, 119 pages.

DM 18,50; approx. US $ 8.60

ISBN 3-540-09609-4

Contents: Background and Scope. – The Tensorial Approach: The Cartesian Tensor Formulation. The Boolean Tensor Formulation. – The Graph Theoretic Approach: The Reachability Criterion. The Term Rank Criterion. – Digraph Decomposition and Tensor Aggregation: On Partitioning of a Digraph. Towards a Total Systems Description. – Some General Remarks. – References.

The knowledge of the separate parts of a system is often not sufficient to describe the system because the interconnections and interactions between the systems elements are of vital importance. An increase in size and complexity calls for new concepts and procedures in the design process of automatic control systems.

This monograph deals with a set of design concepts that qualitatively represents the properties of controllability and observability of time-invariant linear systems. The design rules are formulated in APL.

The authors introduce the concepts of so-called potential controllability and potential observability and derive those concepts from an algebraic as well as a topological point of view. Their main virtue as design tools lies in their organized handling of large numbers of state-variables defining systems of high order or high dimensionality.

Volume 18

Modelling and Optimization of Complex System

Proceedings of the IFIP-TC 7 Working Conference Novosibirsk, USSR, 3–9 July, 1978

Editor: G. I. Marchuk

1979. 59 figures, 5 tables. VI, 293 pages.

DM 32,50; approx. US $ 15.20

ISBN 3-540-09612-4

Contents: Invited Speakers: *A. Bertuzzi, C. Bruni, A. Gandolfi, G. Koch:* Maximum Likelihood Identification of an Immune Response Model. *M. Jilek, P. Klein:* Stochastic Model of the Immune Response. *A. Kalliauer:* A Computational Method for Hierarchical Optimization of Complex Systems. *K. D. Malanowski:* On Regularity of Solutions to Constrained Optimal Control Problems for Systems with Control Appearing Linearly. *R. R. Mohler:* Bilinear Control Structures in Immunology. – **Mathematical Models in Immunology:** *A. L. Asachenkov, G. I. Marchuk:* Mathematical Model of a Disease and Some Results of Numerical Experiments. *L. N. Belykh, G. I. Marchuk:* Chronic Forms of a Disease and Their Treatment according to Mathematical Immune Response Models. *B. F. Dirbov, M. A. Livshits, M. V. Volkenstein:* The Effect of a Time Lag in the Immune Reaction. *E. V. Gruntenko:* Immunological Aspects of Neoplastic Growth. *V. P. Lozovoy, S. M. Shergin:* Structure-Functional Arrangement of the Immune System. *G. I. Marchuk:* Mathematical

Models in Immunology and Their Interpretations. *R. V. Petrov:* Clinical Immunology: Problems and Prospects. – **Optimization of Complex Systems:** *V. M. Alexandrov:* Approximate Solution of Optimal Control Problems. *K. Bellmann, J. Born:* Numerical Solution of Adaptation Problems by means of an Evolution Strategy. *V. L. Beresnev, E. KH. Gimadi, V. T. Dementiev:* Some Models of Taking Optimal Decisions in Standardization. *J. Dolezal:* On Optimal Control of Discrete Systems with Delays. *Yu. P. Drobyshev, V. V. Pukhov:* Analysis of the Influence of a System on Objects as a Problem of Transformation of Data Tables. *E. V. Iljin, A. V. Medvedev, N. F. Novikov:* On Non-Parametric Algorithms of Optimization. *M. Lucertini, A. Marchetti Spaccamela:* Approximate Solutions of an Integer Linear Programming Problem with Resource Variations. *V. M. Matrosov, S. N. Vassilyev, O. G. Divakov, A. I. Tyatushkin:* On Technology of Modelling and Optimization of Complex Systems. *G. I. Marchuk, V. V. Penenko:* Application of Optimization Methods to the Problem of Mathematical Simulation of Atmospheric Processes and Environment. *A. Marrocco, O. Pironneau:* Optimum Design of a Magnet with Finite Elements. *A. Maslowski:* Finite Elements Approximation in State Identification of Large Scale Linear Space-Distributed Systems. *A. A. Petrov, I. G. Pospelov:* Mathematical Modelling of Socio-Economic System. *V. M. Yakovlev:* On one Representation of Necessary Conditions of Optimality for Discrete Controlled Systems.

Volume 19

Global and Large Scale System Models

Proceedings of the Center for Advanced Studies (CAS) International Summer Seminar, Dubrovnik, Yugoslavia, August 21–26, 1978

Sponsored by: The Faculty of Organizational Sciences, University of Belgrade, Yugoslavia

Editor: B. Lazarević

1979. 36 figures, 4 tables. VIII, 232 pages.

DM 28,–; approx. US $ 13.10

ISBN 3-540-09637-X

Contents: *A. V. Balakrishnan:* Seminar on Global and Large Scale System Models – Summary and Conclusions. *R. E. Kalman:* A System-Theoretic Critique of Dynamic Economic Models. *R. Tomović:* Modelling of Large Systems. *A. V. Balakrishnan:* System Science Methodology in Global Modelling: System Identification. *M. Mesarovic:* Practical Application of Global Modelling. *H. D. Scolnik:* A Critical Review of Some Global Models. *A. N. Elshafei:* Global Modelling: Survey and Anticipation of Future Progress. *S. V. Dubovsky, G. G. Pirogov:* Critical Survey of Some Global Modelling Approaches. *G. A. Albizuri:* What Future (If Any) May Global Modelling Have? *S. Holly:* The Construction and Optimal Control of Econometric Models of National Economies for Forecasting and Policy Analysis. *L. W. Taylor, Jr.:* A Low Order Systems Model of the United States Economy. *K. D. Tocher:* The Automatic Translation of Real World Problems into a Mathematical Programming Formulation.

Volume 20

B. Egardt

Stability of Adaptive Controllers

1979. 16 figures, 2 tables. V, 158 pages.

DM 23,–; approx. US $ 10.70

ISBN 3-540-09646-9

Contents: Introduction. – Unified Description of Discrete Time Controllers: Design method for known plants. Class of adaptive controllers. Example of the general control scheme. The positive real condition. – Unified Description of Continuous Time Controllers: Design method for known plants. Class of adaptive controllers. Examples of the general control scheme. The positive real condition. – Stability of Discrete Time Controllers: Preliminaries. L^∞-stability. Convergence in the disturbance-free case. Results on other configurations. Discussion. – Stability of Continuous Time Controllers: Preliminaries. L^∞-stability. Convergence in the disturbance-free case. – References. – Appendices.

Adaptive control is one possibility of tuning controllers. In particular, self-tuning regulators (STR) and model reference adaptive systems (MRAS) are two widely discussed approaches to solve the problem for systems with unknown parameters. These techniques are the main concern of the present work, where several MRAS and STR are described as special cases of a fairly general algorithm. A further topic is the stability analysis of adaptive schemes in the presence of disturbances.

Volume 21
M. B. Zarrop

Optimal Experiment Design for Dynamic System Identification

1979. 9 figures, 4 tables. X, 197 pages.
DM 28,–; approx. US $ 13.10
ISBN 3-540-09841-0

Contents: Preliminaries. – Problem Statement. – A Tchebycheff System Approach. – D-Optimal Designs. – Continuous-Time Systems. – Sampling Rate Design. – Conclusions and Further Research. – References. – Main Appendix: Tchebycheff System Theory.

This volume covers the problem of experiment design for the efficient identification of a linear single input-single output dynamic system from input-output data even in the presence of disturbances. Certain factors can be selected in order to maximize information from an experiment.
A geometrical approach to the design problem is developed for both continuous-time and discrete-time systems. Conditions are derived for the existence of certain minimal representations of the optimal input spectrum, leading to a reduction in the dimension of the design optimization problem.

Volume 22

Optimization Techniques

Proceedings of the 9th IFIP Conference on Optimization Techniques Warsaw, September 4–8, 1979

Part 1

Editors: K. Iracki, K. Malanowski, S. Walukiewicz
1980. 12 figures, 5 tables. XVI, 569 pages.
DM 59,–; approx. US $ 27.50
ISBN 3-540-10080-6

Contents: Plenary Lectures: *A. V. Balakrishnan:* Optimal Control Problems in Aeroelasticity. *W. Gutkowski:* Optimization of Engineering Structures. *L. Kantorovich:* Mathematic-Economic Modelling of Scientifical and Technical Progress. *N. N. Krasovskii:* Game-Theoretical Optimization of Differential Systems. *R. Kulikowski:* Optimization of Regional Development – Integrated Models for Socio-Economic and Environmental Planning. *J. L. Lions:* On the Fundations of the Optimal Control of Distributed Systems. *M. J. D. Powell:* Optimization Algorithms in 1979. *A. P. Wierzbicki:* A Methodological Guide to Multiobjective Optimization. – **Round Table Session on Systems Techniques in Economics. Panel Addresses:** *M. D. Intriligator:* Interactions Between Economics and Systems Theory. *R. F. Drenick:* Modeling Man in Economics and System Theory. *J. M. Robinson:* The Global 2000 Study: An Attempt to Increase Consistency in Government Forcasting. – **Stochastic Control:** *V. P. Belavkin:* Optimization of Quantum Observation and Control. *N. Christopeit:* The Limited Risk Problem in the Nonlinear Case. *J. Dupačová:* Minimax Stochastic Programs with Nonseparable Penalties. *O. Hernández-Lerma:* Exit Probabilities for Degenerate Systems. *W. Römisch:* An Approximation Method in Stochastic Optimal Control. *Ł. Stettner, J. Zabczyk:* Stochastic Version of a Penalty Method. *K. Uchida, E. Shimemura:* On a Class of Linear-Quadratic Stochastic Team Control Problems. – **Differential Games:** *T. Basar:* On the Existence an Uniqueness of Closed-Loop Sampled-Data Nash Controls in Linear-Quadratic Stochastic Differential Games. *A. J. de Zeeuw:* Stackelberg Solutions in Macroeconometric Policy Models with a Decentralized Decision Structure. *J. Doležal:* Differential Games with Parameters. *A. Idzik:* Two-Stage Noncooperative Stochastic Games with Denumerable State Spaces. *M. Medveď:* On a Nonlinear Evasion Problem Described by a System of Integro-Differential Equations. *K. Mizukami, V. Tews:* State-Estimation in a Pursuit-Evasion-Game with Incomplete Information-Exchange. *Gy. Sonnevend:* Existence and Numerical Computation of Extremal Invariant Sets in Linear Differential Games with Bounded Controls. *B. Tolwinski:* Closed-Loop Stackelberg Solution and Threats in Dynamic Games. – **Optimal Control: Ordinary and Delay Differential Equations:** *V. M. Alexandrov:* Quasioptimal Control and Stabilization at Random Perturbations. *P. Brunovsky:* Regular Synthesis and Singular Extremals. *M. Buric, E. B. Lee:* A Tensor Algebraic Approach to Optimal Synthesis for Nonlinear Systems. *W. W. Hager,*

G. D. Ianculescu: Semi-Dual Approximations in Optimal Control. *B. Jakubczyk, B. Kaśkosz:* Realizations of Volterra Series. *R. Klötzler:* Applications of a General Duality Conception in Optimal Control. *J. Komornik:* Asymptotic Behavior of Solutions of Nonautonomeous Riccati Equations. *F. Lhote, B. Lang, J. C. Miellou, P. Spiteri:* Relaxation Methods for Parallel in Line Calculations of the Optimum Control of Large Systems. *B. E. A. Milani:* On the Computation of the Optimal Constant Output Feedback Gains for Large-Scale Linear Time-Invariant Systems Subjected to Control Structure Constraints. *A. W. Olbrot, A. Sikora:* On Sensitivity Minimization for Linear Control System. *B. L. Pierson, I. Chen:* Minimum-Time Soaring Through a Specified Wind Distribution. *T. Zolezzi:* A Variational Characterization of Linear Control System. – **Optimal Control: Partial Differential Equations:** *A. B. Bakushinsky:* Some Results in Approximate Methods for Variational Inequalities with Applications. *E. Brietzke, P. Novosad:* Optimal Control of Eigenvalues – I. *C. Cuvelier:* A Free Boundary Problem in Hydrodynamic Lubrication Governed by the Stokes Equations. *A. Favini:* On Stabilizability of Some Abstract Degenerate Diffusion Processes. *T. L. Johnson:* Optimization of Low-Order Compensators for Infinite-Dimensional Systems. *U. Mackenroth:* Some General Considerations on Optimality Conditions for State Constrained Parabolic Control Problems. *M. Niezgódka, I. Pawlow:* Optimal Control for Parabolic Systems with Free Boundaries – Existence of Optimal Controls, Approximation Results. *L. Pandolfi:* Output Stabilization of a Class of Boundary Value Control Systems. *S. A. Pohjolainen, H. N. Koivo:* Properties and Calculation of Transmission Zeros for Distributed Parameter Systems. *S. Raczyński:* On Optimal Control and Reachable Sets in a Banach Spache. *J. Sokołowski:* Control in Coefficients for Parabolic Equation. *L. von Wolfersdorf:* Necessary Optimality Conditions for Optimal Control Problems with Elliptic Systems in the Plane. **Multiobjective Problems:** *A. Bieluszko:* Task Allocation in Two-Level Systems with Conflicting Goals. *J. Majchrzak:* On Relations Between Continuous and Discrete Multicriteria Optimization Problems. *M. Mańas:* Multiple Pay-Off Conflicts. *V. S. Molostvov, V. I. Zhukovskii:* On Optimality in a Class of Cooperative Many Player Differential Games. *A. N. Nonova, S. K. Stoyanov:* A Vector-Valued Criterion Optimization Method. *S. Opricović:* An Extension of Compromise Programming to Solution of Dynamic Multicriteria Problem. – **Applications: Biomedical Systems:** *A. L. Asachenkov, L. N. Belykh:* On Mathematical Modelling of a Disease. *B. D. Kiekebusch-Müller, H. Arnold:* An Iterative Method for Parameter Estimation in Gene – Counting Procedures. *P. Klein, J. Doležal, J. Šterzl:* Mathematical Model of Regulation of Antibody Response. *R. R. Mohler, W. J. Kołodziej:* On Stochastic Control in Immunology. *D. M. Wiberg, O. Brovko:* Some Factors Affecting the Rate of Convergence During On-Line Parameter Identification.

Volume 23

Optimization Techniques

Proceedings of the 9th IFIP Conference on Optimization Techniques Warsaw, September 4–8, 1979

Part 2

Editors: K. Iracki, K. Malanowski, S. Walukiewicz

1980. 92 figures, 36 tables. XV, 621 pages
DM 59,–; approx. US $ 27.50
ISBN 3-540-10081-4

Contents: Mathematical Programming: Theory: *K. H. Elster, R. Nehse:* Optimality Conditions for Some Nonconvex Problems. *B. Gollan:* A General Perturbation Theory for Optimization Problems. *J. Guddat, M. Lips:* On the Theoretical Basis for Methods for Parametric Optimization Problems. **Mathematical Programming: Algorithms:** *E. J. Anderson:* Basic Solutions and a „SIMPLEX" Method for a Class of Continous Linear Programs. *F. Archetti:* A Probabilistic Algorithm for Global Optimization Problems with a Dimensionality Reduction Technique. *K. Beer:* The Method of Feasible Direction for Optimization Problems with Subdifferentiable Objective Function. *L. Grandinetti:* Factorized Variable Metric Algorithms for Unconstrained Optimization. *Ch. Grossmann:* A Unified Approach to Nonlinear Programming Algorithms Basing on Sequential Unconstrained Minimizations. *J. Hald, K. Madsen:* Minimax Optimization Using Quasi-Newton Methods. *R. Hornung:* Algorithm for the Solution of a Discrete Minimax Problem: Subgradient Methods and a New Fast Newton-Method. *G. S. Lbov:* Algorithm of Search for Global Extremum of Function from Variables Measured in Different Scales. *G. di Pillo, L. Grippo, F. Lampariello:* A

Method for Solving Equality Constrained Optimization Problems by Unconstrained Minimization. *K. Schittkowski:* Randomly Generated Nonlinear Programming Test Problems. *J. S. Sosnowski:* Method of Regularized Approximations and its Application to Convex Programming. *P. Tatjewski:* Methods of Hierarchical Optimization for Interconnected Systems. *E. Toczyłowski:* Structural Analysis of Large Nonlinear Programming Problems. *A. Žilinskas:* On the Use of Statistical Models of Multimodel Functions for the Construction of the Optimization Algorithms. – **Integer Programming:** *B. Bank:* Stability Analysis in Pure and Mixed-Integer Linear Programming. *P. Bertolazzi, C. Leporelli, M. Lucertini:* Alternative Group Relaxation of Integer Programming Problems. *M. Biró:* Efficient Method Applying Incomplete Ordering for Solving the Binary Knapsack Problem. *P. M. Camerini, F. Maffioli:* Weighted Satisfiability Problems and Some Implications. *U. Derigs:* On Two Methods for Solving the Bottleneck Matching Problem. *G. V. Gens, E. V. Levner:* Fast Approximation Algorithms for Knapsack Type Problems. *D. Hausmann, B. Korte:* Computational Relations Between Various Definitions of Matroids and Independence Systems. *J. Krarup, S. Walukiewicz:* Relations Among Integer Programs. *U. Zimmermann:* Linear Optimization for Linear and Bottleneck Objectives with One Nonlinear Parameters. – **Software Problems:** *J. Bisschop, A. Meeraus:* Selected Aspects of a General Algebraic Modeling Language. *A. Kalliauer:* Software Design for Algorithms of Hierarchical Optimization. *K. Kallio:* Outlines for a General Mathematical Modeling Software. *R. C. Rosenberg, A. N. Andry:* An Efficient Algorithm for Obtaining the Reduced Connection Equations for a Class of Dynamic Systems. *K. Yajima:* Characteristics of Incremental Assignment Method. – **Graphs and Networks:** *M. Bereziński, J. Hołubiec:* Stochastic Modelling of Socio-Economic System. *B. Betrò:* Optimal Allocation of a Seismographic Network by Nonlinear Programming. *Z. Bubnicki, M. Staroswiecki, A. Lebrun:* Stochastic Approach to the Two-Level Optimization of the Complex of Operations. *Ph. Chretienne:* Some Results on Timed Petri-Nets. *P. Hammad, J. M. Raviart:* Non Equilibrium Computer Network Distribution. *E. Höpfinger:* Dynamic Programming of Stochastic Activity Networks with Cycles. *R. W. Liebermann, I. B. Türksen:* A Necessary Condition for the Elimination of Crane Interference. *M. M. Sysło:* Optimal Constructions of Project Network. *C. van Nuffelen:* Enumeration Techniques in Directed Hypergraphs. – **Scheduling and Allocation Problems:** *A. Adamski:* Optimal Dispatching Control of Bus Lines. *L. Bianco, M. Cini, L. Grippo:* A Strategic Approach to Air Traffic Control. *G. E. Haramis:* EDP Project and Computer Equipment Selection by the Use of Linear Programming. *S. H. Hung, J. C. Hung, L. P. Anderson:* Impact of Financing on Optimal R a. D Resource Allocation. *J. Kacprzyk, M. Krawczak:* On an Inexact Transportation Problem. *I. Kaliszewski, M. Libura, H. Misiewicz:* Integer Programming as a Tool for Plant Adjustment Problem. *O. B. G. Madsen:* A Cutting Sequencing Algorithm. *J. W. Mercik:* On a Winning Coalition of the Characteristic Function Game as a Solution of the Resource Allocation Problem. – **Applications: Economics and Econometrics:** *C. Bianchi, G. Calzolari, P. Corsi:* A Package for Analytic Simulation of Econometric Models. *K. Brännäs, A. Westlund:* On the Recursive Estimation of Stochastic and Time-Varying Parameters in Economic Systems. *G. M. Folie, A. M. Ulph:* Computing Equilibria in an Industry Producing an Exhaustible Resource. *M. Karasek:* Optimization of a Country's Trade Policies. *M. Luptáčik:* An Open Input-Output Model with Continuous Substitution Between Primary Factors as a Problem of Geometric Programming. *L. Mathiesen, T. Hansen:* An Equilibrium Model for an Open Economy with Institutional Constraints on Factor Prices. *R. Neck:* Controllability and Observability of Dynamic Economic Systems. *V. Zhiyanov:* The Development of Economic System in Case of Differential Optimization/for One-Sector Dynamic Model. – **Applications: Environmental and Energy Systems:** *H. Baumert, P. Braun, E. Glos, W. D. Müller, G. Stoyan:* Modelling and Computation of Water Quality Problems in River Networks. *M. Denny, M. Fuss, L. Waverman:* An Application of Optimal Control Theory to the Estimation of the Demand for Energy in Canadian Manufacturing Industries. *J. A. Hartog, P. Nijkamp, J. Spronk:* Operational Multiple Goal Models for Large Economic Environmental Models. *L. Kruś, M. Libura, L. Słominski:* Resource Distribution Combinatorial Models in Air Pollution Problems. *K. Leimkühler, G. Egberts:* The Energy Economics of the United Kingdom, the Federal Republic of Germany and Belgium. *K. Ogino:* Decentralized Approach for Electric Generating System Development – Energy Supply-Social Siting Concern Interaction. *J. Pintér:* On a Stochastic Model of Reservoir System Sizing. *L. Schrattenholzer:* An LP Energy Supply Model for World Regions. – **Applications: Industrial Processes:** *F. Archetti, C. Vercellis:* An Application of Nonlinear Programming Techniques to the Energy – Economic Optimization of

Building Design. *R. Homescu:* Optimization of the Signal-to-Noise Ratio in the Optical Processing. *W. Jankowski:* An Asymptotic Approach to the Dynamic Optimization of Complex Cyclic Processes. *V. B. Larin:* Methods of Peridic Optimization in Stabilization Problems of Biped Apparatus. *K. Leiviskä:* Comparison of Optimal and Suboptimal Methods for Pulp Mill Production Control. *A. Podsiadło, J. Sobociński:* Streams of Information in the Process of Systematic Modelling of Complex Technical Objects on the Example of Vessel Engines.

Volume 24

Methods and Applications in Adaptive Control

Proceedings of an International Symposium, Bochum 1980
Editor: H. Unbehauen
1980. 121 figures, 9 tables. VI, 309 pages.
DM 36,50; approx. US $ 17.00
ISBN 3-540-10226-4

Contents: Methods in Adaptive Control: *K. J. Åström:* Design Principles for Self-Tuning Regulators. *B. Wittenmark, K. J. Åström:* Simple Self-Tuning Controllers. *D. Matko:* Some Relations in Discrete Adaptive Control. *M. A. El-Bagoury, M. M. Bayoumi:* Multivariable Self Tuning Augmented Regulator. *Ph. de Larminat:* Unconditional Stabilizers for Nonminimum Phase Systems. *T. Söderström:* On Some Adaptive Controllers for Stochastic Systems with Slow Output Sampling. *R. Breddermann:* Realization and Application of a Self-Tuning On-Off Controller. *K. S. Narendra:* Recent Developments in Adaptive Control. *K. Sobel, H. Kaufman, O. Yekutiel:* Design of Multivariable Adaptive Control Systems without the Need for Parameter Identification. *L. Dugard, I. D. Landau:* Convergence Analysis of M. R. A. S. Schemes Used for Adaptive State Estimation. *H. Elliot:* Non Model Reference Adaptive Model Matching. *A. Kuzucu, A. Roch:* Suboptimal Adaptive Feedback Control of Nonlinear Systems. *J. E. Marshall:* Identification Strategies for Time-Delay Systems. *J. Sternby:* Adaptive Control of Extremum Systems. – **Applications of Adaptive Control:** *P. C. Parks, W. Schaufelberger, Chr. Schmid, H. Unbehauen:* Applications of Adaptive Control Systems. *J. van Amerongen:* Model Reference Adaptive Control Applied to Steering of Ships. *P. P. J. van den Bosch, W. Jongkind:* Model Reference Adaptive Satellite Attitude Control. *E. Irving:* Implicit Reference Model and Optimal Aim Strategy for Electrical Generator Adaptive Control. *P. Bonanomi, G. Güth:* Adapted Regulator for the Excitation of Large Turbogenerators. *A. H. Glattfelder, W. Schaufelberger:* Adaptive Control by Self-Selection – An Application to Hydropower Control. *U. Claussen:* Adaptive Time-Optimal Position Control with Microprocessor. *C. T. Cao:* A Simple Adaptive Concept for the Control of an Industrial Robot. *W. L. Green, D. J. Sanger, B. N. Suresh:* Using the Self-Optimising Control of an Electro-Hydraulic Servo System to Minimise the Power Loss. *E. G. Kunze:* Adaptive Control by a Sensitivity Method without Need for On-Line Identification.

Volume 25

Stochastic Differential Systems

Filtering and Control
Proceedings of the IFIP-WG 7/1 Working Conference Vilnius, Lithuania, USSR, Aug. 28–Sept. 2, 1978
Editor: B. Grigelionis
1980. IX, 363 pages.
DM 41,–; approx. US $ 19.10
ISBN 3-540-10498-4

Contents: *R. Z. Hasminskii, I. A. Ibragimov:* Some Estimation Problems for Stochastic Differential Equations. *A. M. Yaglom:* Applications of Stochastic Differential Equations to the Description of Turbulent Equations. *B. Grigelionis, R. Mikulevicius:* On Semimartingales with Values in Euclidean Halfspaces. *Yu. L. Dalecky:* Multiplicative Operator Functional of Markov Processes and Their Applications. *L. I. Galtchouk:* On the Predictable Jumps of Martingales. *V. A. Lebedev:* On the Existence of a Solution of the Stochastic Equation with Respect to a Martingale and a Random Measure. *H. Pragarauskas:* On Bellman Equation for Controlled Degenerated General Stochastic Processes. *A. Yu. Veretennikov:* On the Existence of the Optimal Policy for a Multidimensional Quasidiffusion Controlled Process. *D. Vermes:* On the Semigroup Theory of Stochastic Control. *M. I. Višic, A. I. Komech:* Stationary Solutions of the Stochastic

Navier-Stokes Equations. *Yu. M. Kabanov, R. Sh. Liptser, A. N. Shiryayev:* On Absolute Continuity of Probability Measures for Markov-Itô Processes. *C. Bromley, G. Kallianpur:* Representations of Gaussian Random Fields. *Kiyosi Itô:* Continuous Additive S' Processes. *O. A. Glonti:* Stochastic Differential Equation of the Optimal Non-Linear Filtering of the Conditional Gaussian Process. *J. M. C. Clark, R. J. Cameron:* The Maximum Rate of Convergence of Discrete Approximations for Stochastic Differential Equations. *E. Platen:* Approximation of Itô Integral Equations. *H. J. Engelbert, J. Heß:* A Probabilistic Approach to the Representation Problem of Martingales as Stochastic Integral. *G. C. Papanicolaou, S. R. S. Varadhan:* Diffusion in Regions with Many Small Holes. *M. Cranston, S. Orey, U. Rösler:* Exterior Dirichlet Problems and the Asymptotic Behavior of Diffusions. *A. V. Balakrishnan:* On Stochastic Bang-Bang Control. *H. J. Fischer:* Structure of Martingales under Random Change of Time. *Ya. I. Belopolskaya:* On Stochastic Equations with Unbounded Coefficients for Jump Processes. *B. I. Arkin, M. T. Saksonov:* To the Maximum Principle Theory for Problems of Control of Stochastic Differential Equations. *S. V. Anulova:* Diffusion Processes with Singular Characteristics. *J. M. Stoyanov, O. B. Enchev:* Construction and Properties of a Class of Stochastic Integrals. *A. F. Taraskin:* The Asymptotic Statistical Problems for Fields of Diffusion Type. *B. L. Rozovsky:* A Note on Strong Solutions of Stochastic Differential Equations with Random Coefficients. *H. Rost:* Non-Equilibrium Solutions of an Infinite System of Stochastic Differential Equations. *A. A. Novikov:* On Conditions for Uniform Integrability for Continuous Exponential Martingales. *R. Morkvenas:* On Weak Compactness of the Sets of Multiparameter Stochastic Processes. *S. Ja. Mahno:* Limit Theorems for Stochastic Equations with Partial Derivatives. *V. Mackevičius:* Formula for Conditional Wiener Integrals. *G. L. Kulinič:* On the Asymptotic Behavior of the Solution of the Dimensional Stochastic Diffusion Equation. *V. V. Jurinskii:* On a Dirichlet Problem with Random Coefficients. *V. L. Girko:* Stochastic Spectral Equations.

Volume 26

D. L. Iglehart, G. S. Shedler

Regenerative Simulation of Response Times in Networks of Queues

1980. 22 figures, 17 tables. XII, 204 pages.

DM 28,–; approx. US $ 13.10

ISBN 3-540-09942-5

Contents: Introduction. – Simulation of Regenerative Processes. – Closed Networks of Queues. – The Marked Job Method. – Examples and Simulation Results. – Finite Capacity Open Networks. – Marked Job Simulation Via Hitting Times. – The Decomposition Method. – Efficiency of Simulation. – Networks with Multiple Job Types. – Implementation Considerations. – Appendices. – References. – Index of Notation.

This monograph deals with probabilistic and statistical methods for discrete event simulation of networks of queues. The emphasis is on the use of underlying stochastic structures for the design of simulation experiments and the analysis of simulation output. It focuses on recently developed methods for estimating general characteristics of "passage times" in closed networks of queues. Informally, passage time is the time required for a job to traverse a portion of a network. Such quantities are important in computer and communication system models where they represent job response times. In this context, expected values as well as other characteristics of response times are of interest.
The material is accessible to simulation practitioners as well as to students and researchers interested in the methodology of discrete event simulation. Some knowledge of elementary probability theory, statistics, and stochastic models is sufficient to understand the estimation procedures and the examples.

Volume 27

Extensions of Linear-Quadratic Control Theory

By D. H. Jacobson, D. H. Martin, M. Pachter, T. Geveci

1980. 6 figures. XI, 288 pages.

DM 32,50; approx. US $ 15.20
ISBN 3-540-10069-5

Contents: Review of Linear Dynamic Systems. – Canonical Forms, Pole Assignment and State Observers. – Lyapunov Stability Theory. – Linear-Quadratic Optimal Control. – Introduction to Kalman Filtering. – The Maximum Principle and the Hamilton-Jacobi-Bellman Equation. – The Non-Convex Case. – Controllability Subject to Controller Constraints. – Linear-Quadratic Problems with Conical Control Set. – Special Non-Linear-Quadratic Formulations. – Hybrid Criteria and Parameter Sensitivity. – Conditional Definiteness of Quadratic Functionals. – Exponential Performance Criteria and State-Dependent Noise. – Differential Games. – Optimal Control of Partial Differential Equations.

This volume is based on a series of lectures and research work at the National Research Institute for Mathematical Sciences (NRIMS) of the Council for Scientific and Industrial Research in Pretoria.
It is a concise survey of much of the theory of linear-quadratic optimal control initiated by R. E. Kalman around 1960. It also covers details of various extensions to the theory with recent research results. Some familarity with real vectors and matrices, with systems of ordinary differential equations and with some basic Hilbert Space theory is presupposed; the flavor, however, is decidedly applied mathematical, and these notes should be of use principally to those concerned with process control in industry, dynamic modelling and decision theory in economics and management science, and to members of departments of applied mathematics and engineering at universities and technical colleges.

Volume 28

Analysis and Optimization of Systems

Proceedings of the Fourth International Conference on Analysis and Optimization of Systems
Versailles, December 16–19, 1980
Editors: A. Bensoussan, J. L. Lions
1980. 167 figures, 23 tables. XIV, 999 pages (206 pages in French).
DM 98,–; approx. US $ 45.60
ISBN 3-540-10472-0

Contents: Table of Contents – Table des Matières: I – Large Scale Systems/Grands Systèmes: *P. V. Kokotovic, R. G. Phillips, S. H. Javid:* Singular Perturbation Modeling of Markov Processes. *A. Mori:* Computer Control Systems of a Bloom Caster. *C. Gomez, P. Ratte, C. Saguez, F. Versini:* Simulation d'un train finisseur de laminage à chaud. *S. A. Pohjolainen:* Robust Multivariable Pi-Controller for Distributed Parameter Systems. *B. Durand, G. Cohen, C. Fondraz:* Optimization et centralisation de la gestion de l'énergie sur un site industriel complexe. – **II – Multivariable Systems/Systèmes Multivariables:** M. C. Delfour: Status of the State Space Theory of Linear Hereditary Differential Systems with Delays in State and Control Variables. *A. Manitius:* Algebraic Criteria of Approximate Controllability of Retarded Systems. *A. C. Antoulas:* A Polynomial-Matrix Solution of the Disturbance Localization Problem. *J. P. Gauthier, G. Bornard:* Stabilization of Bilinear Systems, Performance Specification and Optimality. *J. Starr, M. Wemans, J. Vandewalle:* Comparison of Multivariable MBH realization Algorithms in the Presence of Multiple Poles, and Noise Disturbing the Markov Sequence. – **III – Adaptive Systems/Systèmes Adaptatifs:** *D. D. Falconer:* Adaptive Filter Theory and Applications. *E. Mosca, G. Zappa:* MUSMAR: Basic convergence and Consistency Properties. *J. J. Fuchs:* Commande Adaptative Explicite – un Exemple. *R. J. Evans, G. C. Goodwin, R. Betz:* Discrete Time Adaptive Control for Classes of Nonlinear Systems. *E. Irving, H. Dang van Mien:* Discrete Time Model Reference Multivariable Adaptive Control. Applications to Electrical Power Plants. – **IV – Stochastic Dynamical Systems/Systèmes Dynamiques Stochastiques:** *M. H. A. Davis, P. H. Wellings:* Computational Problems in Nonlinear Filtering. *S. K. Mitter:* Recent Results on Nonlinear Filtering. *H. F. Chen:* Least Squares Identification for Continuous Time Systems. *M. Pavon:* On the Gohberg-Krein Factorization and the Conjugate Process. *G. Mazziotto, J. Szpirglas:* Théorème de séparation pour le contrôle impulsionnel: cas Markovien à espace d'état fini. *W. H. Fleming:* Stochastic Control under Partial Observations. *M. Pavon, R. J. B. Wets:* A Stochastic Varational Approach to the Duality Between Estimation and Control: Continuous Time. *F. Delebecque, J. P. Quadrat:* The Optimal Cost Expansion of Finite Controls Finite States Markov Chains with Weak and Strong Interactions. *J. L. Menaldi:* On Degenerate Varational and Quasi-Varational Inequalities of Parabolic Type. – **V – Games, Theory and Applications/ Jeux, Thèorie et Applications:** *Y. C. Ho, P. B.*

Luh, G. J. Olsder: A Control-Theoretic View on Incentives. *J. B. Cruz, Jr.:* Survey of Leader Follower Concepts in Hierarchical Decision Making. *T. Başar:* Memory Strategies and a General Theory for Stackelberg Games with Partial State Information. *G. Huberman:* The Nucleolus and the Essential Coalitions. *F. Angrand, B. Enjalbert, M. H. Fouche, C. Lemarechal:* Gradient Type Optimization Methods to Solve Differential Games Applied to Tri-Dimensional Air-to-Air Combats. – **VI – Distributed Parameter Systems. Theory and Applications – Systèmes à Parametrès Distribués, Théorie et Applications:** *J. P. Yvon:* Contrôle optimal et sous-optimal des systèmes distribués. *S. G. Tzafestas:* Multilevel Stackelberg Control of Distributed-Parameter Systems. *M. Courdesses, A. Martinez, M. Amouroux:* Modelisation et identification paramétrique d'un processus de diffusion assistée par implantation de protons. *I. Derese, E. Noldus:* Control of Parallel Current and Countercurrent Heat Exchangers. *A. Munack:* Application of Adaptive Control to a Bubble-Column-Fermenter. *T. I. Seidman:* Regularity of Optimal Boundary Controls for Parabolic Equations. – **VII – Algebraic and Geometric System Theory – Théorie Algébrique et Geométrique des Systèmes:** *R. E. Kalman:* Nonlinear Realization Theory. *J. C. Willems:* Almost Noninteracting Control Design Using Dynamic State Feedback. *M. L. J. Hautus, M. Heymann:* New Results on Linear Feedback Decoupling. *C. Lobry:* Cycles Limites et Boucles de Rétroaction. *M. Fliess, D. Normand-Cyrot:* Vers une approche algébrique des systèmes non linéaires en temps discret. – **VIII – Economic Systems/Systèmes Economiques:** *M. D. Intriligator:* The Applications of Control Theory to Economics. *E. Burmeister, K. D. Wall:* Estimation of Unobserved Rational Expectations with an Application to the German Hyperinflation. *B. Rustem:* Policy Optimization Algorithms for Nonlinear Econometric Models. *G. Ricci:* Adaptive Control of Linear Decentralized Econometric Models. *T. F. Cooley:* Recursive Estimation of Price Expectations in Economics. – **IX – Multidimensional Systems and Applications to Image Processing/Systèmes Multidimensionnels et Applications au Traitement d'Images:** *L. M. Silverman, F. J. Clara:* Recent Results in Recursive and Nonlinear Image Restoration. *R. M. Mersereau, Th. C. Speake:* Generalized Cooley-Tukey Algorithms for Evaluation of Multidimensional Discrete Fourier Transforms. *P. N. Paraskevopoulos:* Feedback Design Techniques for Linear Multivariable 2-D Systems. *Ph. Delsarte, Y. Genin, Y. Kamp:* Stability of Multidimensional Systems. *O. D. Faugeras:* Optimization Techniques in Image Analysis. – **X – Modelling of Oil Fields / Modelisation des Champs Pétrolifères:** *C. Anderson, P. Concus:* A Stochastic Method for Modeling Fluid Displacement in Petroleum Reservoirs. *B. Dupraz, M. Latil, P. Lemonnier:* Ajustement automatique de modèles de gisements pétroliers: application à l'interprétation d'essais d'interférences sur le gisement de Chuelles. *J. Jaffre:* Simulation numérique de déplacements bidimensionnels d'huile par de l'eau. *F. J. Casse, D. Waldren, M. J. Wheatley, A. Settari:* Some Selected Topics in Research and Applications of Reservoir Simulation. – **XI – Applications of Microprocessors to Control/Applications des Microprocesseurs au Contrôle:** *G. Schmidt:* The Role of Multi-Microcomputers in Automatic Control. *I. Cohen, R. Hanus, D. van Laethem:* Nouvelle méthodologie de l'instrumentation à microcalculateurs destinée à améliorer la conduite automatique des procédés industriels. *A. Halme:* A Two Level Realization of Self-Tuning Regulator in a Multi-Micro-Computer Process Control System. *T. Ezzat:* Online Control of Immobilized Enzyme Reactors. A Microprocessor Implementation. *R. Benejean:* Adaptive Control of Generator Voltage Using Micro-Computers Simulation. *J. M. Dumas, F. Prunet:* C. A. O. de l'implantation optimale de programmes de commande.

Volume 29
M. Vidyasagar

Input-Output Analysis of Large-Scale Interconnected Systems

Decomposition, Well-Posedness and Stability

1981. VI, 221 pages.
DM 28,–; approx. US $ 13.10
ISBN 3-540-10501-8

Contents: Introduction. – Mathematical Preliminaries. – Gain and Dissipativity. – Decomposition of Large-Scale Interconnected Systems. – Well-Posedness of Large-Scale Interconnected Systems. – Small-Gain Type Criteria for L_p-Stability. – Dissipativity-Type Criteria for L_2-Stability. – L_2-Instability Criteria. – L_∞-Stability and L_∞-Instability Using Exponential Weighting. - References. - Index.

This monograph contains a fairly comprehensive review of large-scale interconnected systems from an input-output viewpoint. Prior to treating the question of stability (and instability), both the decomposition and the well-posedness of such systems are studied. It is not necessary for the reader to have studied feedback stability before tackling this book as results concerning feedback systems are developed as special cases of more general results pertaining to large-scale systems. However, the reader should know some elementary functional analysis (e.g. Lebesque spaces, contraction mapping theorem), and have more general knowledge (e.g. Perron-Frobenius theorem). The first chapter is introductory, and chapters 2 and 3 contain background material; the remaining chapters are essentially independent and can be read in any order.

Volume 30

Optimization and Optimal Control

Proceedings of a Conference Held at Oberwolfach, March 16–22, 1980

Editors: A. Auslender, W. Oettli, J. Stoer

1981. VIII, 254 pages.
DM 32,50; approx. US $ 15.20
ISBN 3-540-10627-8

Contents: Part 1: Optimization: *M. Atteia, A. El Qortobi:* Quasi-Convex Duality. *J.-P. Crouzeix:* Some Differentiability Properties of Quasiconvex Functions on $\mathbb{R}^n$. *J. Gwinner:* On Optimality Conditions for Infinite Programs. *J.-B. Hiriart-Urruty:* Optimality Conditions for Discrete Nonlinear Norm-Approximation Problems. *P. Huard:* Feasible Variable Metric Method for Nonlinearly Constrained Problems. *B. Korte, R. Schrader:* A Note on Convergence Proofs for Shor-Khachian-Methods. *C. Lemaréchal:* A View of Line-Searches. *E. A. Nurminski:* Π-Approximation and Decomposition of Large-Scale Problems. *J.-P. Penot:* On the Existence of LAGRANGE Multipliers in Nonlinear Programming in Banach Spaces. *S. Schaible, I. Zang:* Convexifiable Pseudoconvex and Strictly Pseudoconvex C^2-Functions (Extended Abstract). *K. Schittkowski:* Organization, Test, and Performance of Optimization Programs. *P. Spellucci:* Han's Method without Solving QP. **Part 2: Optimal Control:** *M. Brokate:* Necessary Optimality Conditions for Differential Games with Transition Surfaces. *F. Colonius:* Regularization of LAGRANGE Multipliers for Time Delay Systems with Fixes Final State. *W. Hackbusch:* Numerical Solution of Linear and Nonlinear Parabolic Control Problems. *M. Kohlmann:* Survey on Existence Results in Nonlinear Optimal Stochastic Control of Semimartingales. *W. Krabs:* Time-Minimal Controllability in the View of Optimization. *D. Kraft:* On the Choice of Minimization Algorithms in Parametric Optimal Control Problems. *U. Mackenroth:* Strong Duality, Weak Duality and Penalization for a State Constrained Parabolic Control Problem. *K. Malanowski:* Finite Difference Approximations to Constrained Optimal Control Problems.

Volume 31
B. Rustem

Projection Methods in Constrained Optimisation and Applications to Optimal Policy Decisions

1981. XV, 315 pages.
DM 36,50; approx. US $ 17.00
ISBN 3-540-10646-4

Contents: Nonlinearly Constrained Optimisation Techniques Based on Projections. – Projection Methods for Computing Feasible Points of Linearly Constrained Regions. – An Algorithm for Positive Definite Quadratic Programming. – Optimisation with Linear and Nonlinear Constraints. – The Iterative Specification of Objective Functions in Economic Policy Optimisation: An Application of Projection Methods. – Policy Optimisation Algorithms for Nonlinear Econometric Models. – Suggestions for Further Research. – References for Chapters 1–7.

Projection methods in constrained optimisation and the application of projection techniques to policy optimisation problems are the primary topics treated in this work. The constrained optimisation problem of minimizing a nonlinear function of n variables subject to equality and inequality constraints is also known as the nonlinear programming problem. Projection methods of constrained minimization involve projections of descent directions. The basic idea underlying these methods is the principle of projections which may be considered to be the generalisation of the fact that in n-dimensional Euclidean space the shortest vector from a point to a subspace is orthogonal to the subspace.

The work consists of two parts. The first part is concerned with the development of projection techniques for different aspects of constrained optimisation. The second part covers with the application of projection techniques to computational problems in policy optimisation.

Volume 32
T. Matsuo

Realization Theory of Continuous-Time Dynamical Systems

1981. VI, 329 pages.
DM 36,50; approx. US $ 17.00
ISBN 3-540-10682-0

Contents: Introduction. – Realization Theory of (General) Dynamical Systems. – Realization Theory of Linear Representation Systems. – Realization Theory of (Algebraic) Linear (Time-Constant) Systems. – References. – Index of Notations. – Index of Subjects.

This monograph presents a basis of realization theory of dynamical systems. Realization problem is to determine an intrinsic mathematical model (canonical dynamical system) from the input-output relations of a given causal black box, i. e., to understand fully the internal behavior of the black box from the experimental data of it. Morever, if possible, the author wants to make the mathematical model real. Realization theory was mainly developed in the field of discrete-time linear systems as the algebraic system theory. On the other hand, it has been recognized that the highly developed automata theory has close relations with realization theory. Chapter A contains the introduction; chapter B shows that for any input response map a, there is a uniquely determined (up to isomorphisms) canonical dynamical system such that its behavior is a; in chapter C the author considers linear representation systems which are special dynamical systems such that the state spaces have linear structure; in chapter D linear concatenation monoids and (algebraic) linear time-constant systems are defined.

This book may be considered as the continuous-time version of the author's lecture "Realization theory of discrete-time systems" at the graduate course in the School of Engineering, Nagoya University, 1976–1980.

Volume 33
P. Dransfield

Hydraulic Control Systems – Design and Analysis of Their Dynamics

1981. VII, 227 pages.
DM 28,–; approx. US $ 13.10
ISBN 3-540-10890-4

Contents: Introduction. – Notation and Units. – Conventional Modelling Procedures. – Power Flow Modelling. – Power Bond Graphs. – Solution of Power Flow Models. – Selecting Equations and Coefficients. – Applications of Bond Graphs. – Optimizing Dynamic Response. – Phenomena Which Can Affect Response. – Conclusion. – Appendix 1: Static Design Approaches. – Appendix 2: SI Conversion Factors. – Appendix 3: Dynamic Response – A Bibliography. – Index.

For hydraulic control systems the use of power bond graphs is a natural and superior approach to the development of reliable dynamic models. They are introduced, explained, demonstrated, and utilized in substantial detail in this volume.

The monograph is written mainly for those mechanical and control engineers seeking to include predictive dynamic analysis among their techniques for the design of hydraulic control systems and who find existing approaches unsatisfying or inappropriate. The text is also suitable for graduate or senior undergraduate courses, and for specialized extension-type courses.

Volume 34
H. W. Knobloch

Higher Order Necessary Conditions in Optimal Control Theory

1981. V, 173 pages.
DM 23,–; approx. US $ 10.70
ISBN 3-540-10985-4

Contents: Introduction. – **Part I: The formal framework:** Variable switching points. The basic definitions. The formula for $(D\hat{x})(w, a)$ in case $D = \partial^{\upsilon}/(\partial w^{i})^{\upsilon}$. Dependence upon the f_i. Differentiation with respect to w. Linear differential equations. Control systems. Formal Taylor expansions; The general composition rule. Control constraints; Higher order necessary conditions. Auxiliary results. Proof of Theorem 9.1. – **Part II: Interior controls. Second order conditions:** Linear and quadratic approximation of $h^{(\upsilon)}$. The operator Γ. Linear and quadratic approximation of K_{υ}. Transformation of the control variable. Some identities involving Lie-brackets and the operator Γ. A general formula for the Lie-brackets $[B_{\mu}^{p}, B_{\nu}^{q}]$. A special case of the Clebsch-Legendre condition. Auxiliary results. The generalized Clebsch-Legendre condition. Multivariable controls; Second order equality conditions. – **Part III: Free-endpoint problems:** Restatement of previous results. The Jacobson-Gabasov condition and its generalization. – Appendix. – References.

This volume deals with problems in deterministic control theory which originate from the occurrence of "singular" arcs. In general terms, these are portions of optimal solutions for which the bangbang principle does not hold. Since standard optimization techniques often fail to provide satisfactory information about this special type of solution, a variety of local tests which supplement the maximum principle along a singular arc have been proposed within the last 20 years. The monograph provides a complete and self-contained approach to these "higher order" necessary conditions. Except for some elementary facts from optimization theory no more mathematical background is required than advanced calculus and linear algebra.

Volume 35

Global Modelling

Proceedings of the IFIP-WG 7/1 Working Conference Dubrovnik, Yugoslavia, Sept. 1–5, 1980
Editor: S. Krčevinac
1981. Approx. 240 pages.
DM 37,–; approx. US $ 17.20
ISBN 3-540-11037-2

Contents: *R. Tomović:* Top-Down Approach to Global Dynamic Modelling. – *Z. Zarić:* Experience with Energy System Modelling in Serbia. – *Y. R. Cho, E. D. Gahan:* The UNIDO World Industry Co-operation Model. – *S. I. Gass, S. C. Parikh:* Credible Baseline Analysis for Multi-Model Public Policy Studies. – *H. P. Smit, R. P. Vos, H. J. W. Weyland:* Medium- and Long-Term Models for the ESCAP Region. – *H. Bossel:* Modelling Policy Consequences and Evaluation Processes Using the "DEDUC" Nonnumerical Program System. – A. H. Zemanian: Economic Models of Periodic Marketing Systems. – *Y. Zahavi:* A New Urban Travel Model. – *P. M. Allen:* Modelling the Self-Organization of Human Systems. – *S. K. Gehrecke:* The Impact of Energy and Environmental Policy on the Design of Energy Models. – *N. Fergany:* Treatment of the Arab Region in Global Models. – *N. Fergany:* Treatment of the Arab Region in Global Models. – *G. Bruckmann:* IIASA's Role in Global Modelling. – *V. B. Bajić and B. J. Petrović:* Quasi-Models of Price Evolution and their Qualitative Properties. – *G. G. Pirogov:* A Model of Regional Interactions Considering Energy Deficits. – *M. Rajkov, S. Andrić, Z. Šišarica, Lj. Rajin:* Contribution to the Simulation Modelling of an Economic System.

Volume 36

Stochastic Differential Systems

Proceedings of the 3rd IFIP-WG 7/1 Working Conference Visegrád, Hungary, September 15–20, 1980
Editors: M. Arató, D. Vermes, A. V. Balakrishnan
1981. Approx. 250 pages.
DM 32,50; approx. US $ 15.20
ISBN 3-540-11038-0

Contents: *M. Arató:* On optimal stopping times in operating systems. – E. Çinlar, J. Jacod: Semimartingales defined on Markov processes. – *M. A. H. Dempster:* The expected value of perfect information in the optimal evolution of stochastic systems. – *M. D. Donsker, S. R. S. Varadhan:* Some problems of large deviations. – *H. J. Engelbert, W. Schmidt:* On the behaviour of certain functionals of the Wiener process and applications to stochastic differential equations. – *P. Greenwood:* Point processes and system lifetimes. – *B. Grigelionis, R. Mikulevičius:* On weak convergence of semimartingales and point processes. – *I. Gyöngy:* Ito formula in Banach spaces. – *D. I. Hadžiev:* General theorems of filtering with point process observations. – *U. G. Haussmann:* Existence of partially observable stochastic optimal controls. – *S. Ishak, J. Mogyoródi:* On the generalization of the Fefferman-Garsia inequality. – *G. Kallianpur:* Some remarks on the purely nondeterministic property of second order random fields. – *P. Kotelenez:* The Hölder continuity of Hilbert space valued stochastic integrals with an application to SPDE. – *N. V. Krylov, B. L. Rozovsky:* On the first integrals and Liouville equations for diffusion processes. – *H. J. Kushner:* An averaging method for the analysis of adaptive systems with small adjustment rate. – *M. Metivier:* A-spaces associated with processes. Application to stochastic equations. – *A. A. Novikov:* A martingale approach to first passage problems and a new condition for Wald's identity. – *E. Platen:* A Taylor formula for semimartingales solving a stochastic equation. – *Gy. Sonnevend:* On optimal sensor location in stochastic differential systems and in their deterministic analogues. – *H. Pragarauskas:* On first order singular Bellman equation. – *J. von Scheidt:* A limit theorem of solutions of stochastic boundary-initial-value problems. – *J. M. Stoyanov, O. B. Enchev:* Stochastic integration with respect to multiparameter Gaussian processes. – *D. Surgailis:* On L^2 and non-L^2 multiple stochastic integration. – *D. Vermes:* Optimal stochastic control under reliability constraints. – *A. A. Ÿuskevič:* On controlled semi-Markov processes with average reward criterion.

The Authors

Numbers printed in italics refer to monographs and those printed upright refer to proceedings volumes whose contents are listed in **Title Description**

Available from:

Order Form

Please order through your bookseller

or Springer for Science, P. O. B. 503, 1970 AM Ijmuiden, Netherlands
or Springer-Verlag Berlin, Heidelberger Platz 3, D-1000 Berlin 33
or Springer-Verlag New York Inc., 175 Fifth Avenue, New York, NY 10010, USA

Lecture Notes in Control and Information Sciences

Volume ____ ISBN ____________________

....

....

....

....

....

I/we would like to place a standing order for
Lecture Notes in Control and Information Sciences
beginning with Volume ____ and for each volume thereafter.
I/we may cancel the order at any time.

Name/Address: ______________________________

____

____

Date/Signature: ______________________________

Prices are subject to change without notice.

Notes

Notes

Notes

Notes

ISBN 978-3-662-23401-3 ISBN 978-3-662-25449-3 (eBook)
DOI 10.1007/978-3-662-25449-3

Only prices quoted in Deutschmark are binding. U.S.Dollar prices are established according to the exchange rate at the time of sales, Dollar prices shown in this catalog are those prevalent at time of issue.

Originally published by Springer-Verlag Berlin Heidelberg New York in 1981

31799/10.81.136. Be.

GPSR Compliance
The European Union's (EU) General Product Safety Regulation (GPSR) is a set of rules that requires consumer products to be safe and our obligations to ensure this.

If you have any concerns about our products, you can contact us on

ProductSafety@springernature.com

In case Publisher is established outside the EU, the EU authorized representative is:

Springer Nature Customer Service Center GmbH
Europaplatz 3
69115 Heidelberg, Germany

www.ingramcontent.com/pod-product-compliance
Ingram Content Group UK Ltd.
Pitfield, Milton Keynes, MK11 3LW, UK
UKHW021930190726
13853UKWH00002B/966

* 9 7 8 3 6 6 2 2 3 4 0 1 3 *